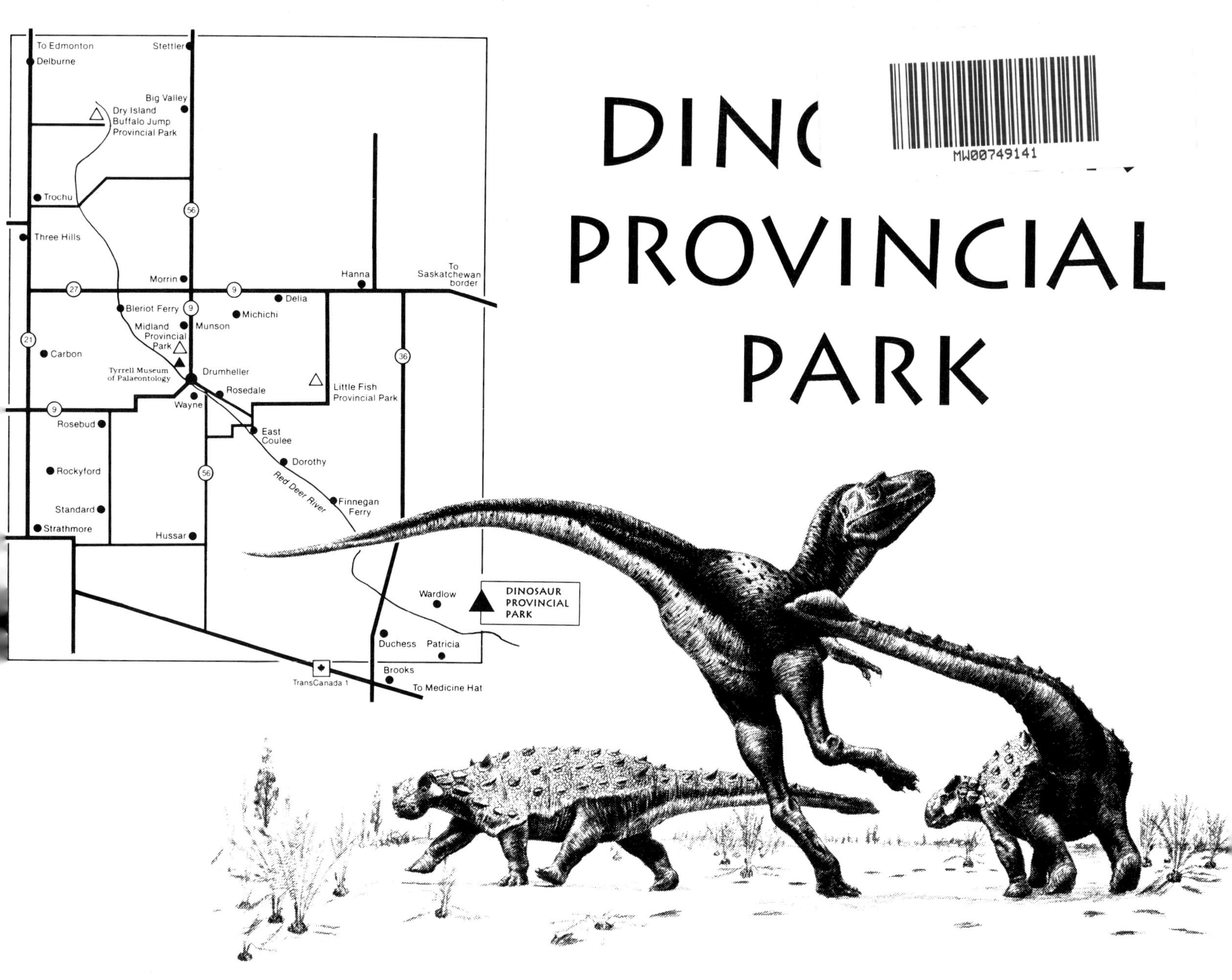



*Albertosaurus vs Euoplocephalus*

*National Museums of Canada*

# Acknowledgements

*I wish to thank the following for their assistance, guidance and informative interviews, especially their time involved between 1983 and 1986:*

*Dr. Loris S. Russell , Professor Rene W. Barendregt, Dr. Philip Currie, Dr. Dale A. Russell; the Tyrrell Museum staff, especially Ann M. Garneau, Joanne Laukulich and Kevin Zak; Dinosaur Provincial Park Head Ranger Roger Benoit and his 1983-85 summer staff; the Canadian Geographical Society (for providing final funding); the Glenbow Museum, Calgary; the Calgary Zoo; Alberta Recreation and Parks; the Lethbridge Public Library; the Smithsonian Institute, Washington D.C.; Richard and Violet Zahn; Brian Preddy; and Neil and Lee Leadbeater.*

*Dr. Philip Currie at the Centrosaurus bone bed.*

# DINOSAUR
# PROVINCIAL PARK

*by Gordon Reid*

Mosaic Press
Oakville-Buffalo-London

*To Brandon, Cameron and Michael and the future of paleontology!*


## Canadian Cataloguing in Publication Data

Reid, Gordon, 1954-
Dinosaur Provincial Park

Rev. ed.
Includes bibliographical references.
ISBN 0-88962-556-5

1. Dinosaur Provincial Park (Alta.). 2. Natural history - Alberta. I. Title.

FC3665.D56R4 1994 971.23'4 C94-931396-3
F1079.D56R4 1994





Published by Mosaic Press, P.O. Box 1032, Oakville, Ontario, L6J 5E9, Canada. Offices and warehouse at 1252 Speers Road, , Units #1&2, Oakville, Ontario L6L 5N9, Canada

Mosaic Press acknowledges the assistance of the Canada Council and the Ontario Arts Council in support of its publishing programme, The Ontario Ministry of Culture, Tourism and Recreation and the Department of Communications, Government of Canada.





Book and Cover Design by Patty Gallinger (Illustrations © P. Gallinger: Pages 5, 7, 8, 11 and 30)

Printed and bound in Canada
ISBN 0-88962-556-5 PB

**MOSAIC PRESS**:
In Canada:
Mosaic Press, 1252 Speers Road, Units 1&2, Oakville, Ontario L6L 5N9,
P.O. Box 1032, Oakville, Ontario, L6J 5E9.
In the U.S.:
Mosaic Press, 85 River Rock Drive, Suite 202, Buffalo, N.Y., 14207,
In the U.K.:
John Calder (Publishers) Ltd., 9-15 Neal Street, London, WCZH 9TU, England.

# Table of Contents



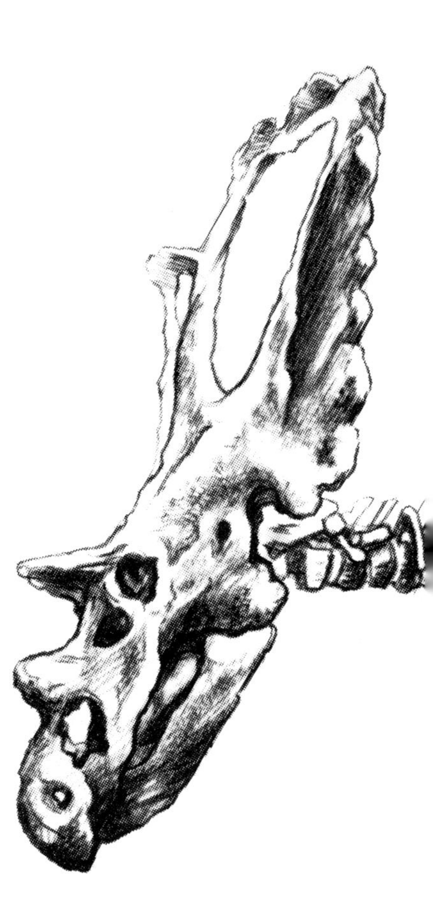

# *Preface*

Dinosaur Provincial Park has gained world-wide notoriety. The park is no longer a "sleeping giant". The Renaissance of Dinosaur Park began with the opening of a field station housing fascinating fossil exhibits indigenous to the area. The new station is like a mini-Royal Tyrrell Museum!

An exchange of scientists from the Ex-Terra Foundation brought Chinese paleontologists to Dinosaur Provincial Park's environs. Numerous television documentaries (featuring Tyrrell's Philip Currie), an on location shoot for "Unforgiven" (Clint Eastwood's latest western) and all the rage of Steven Spielberg's blockbuster film "Jurassic Park" (1993) and now a five year world tour of "The Greatest Show Unearthed" (featuring dinosaur remains from Mongolia) has sent our globe reeling into another all encompassing dinosaur hysteria!

So I felt it was time to update the photographs in *Dinosaur Provincial Park* and hopefully satisfy the curiosity of those who have never been to the beautiful badlands of Alberta.

Although others (notably Renie Gross in *Dinosaur Country)* have written extensively about the badlands, I felt a book dealing directly with Dinosaur Provincial Park was long overdue. Before now there were only assorted diaries, newspaper clippings, various pamphlets and magazine articles.

Thanks to the Canadian Geographical Society's generous grant system, the co-operation of Alberta's Recreation and Parks people and the early encouragement of (geography and geology) Professor Rene W. Barendregt, I was able to gradually explore and photograph numerous areas of this incredible park.

Park naturalists ably assisted me on lengthy hikes in the badlands, carrying equipment and answering questions along the way. Sometimes it got rough. Heat, mosquitoes, descending steep cliff walls, exploring complex piping systems and trekking through a valley at dusk was like something out of the *Twilight Zone.* Manoeuvring through long, deep shadows cast by towering hoodoos, we dodged around dinosaur limb bones and escaped from a curious fruit bat in hot pursuit!

I was impressed overall by the park's hiking trails, which served to expose an unusual vista of spacious landscape. The sudden silence in some areas of the park gripped me with an overwhelming calm, contemplative mood. The stillness tends to fine tune one's

senses and makes one wonder what is over the next hill. It all seems so limitless. As dusk decended, we heard the distant cries of coyotes and the peculiar sounds of diving nighthawks; these sounds still haunt me.

*Dinosaur Provincial Park* opens with a look at the park as it was 70-80 million years ago. A sampling of the dinosaurs indigenous to the park precedes a look at modern theories and paleontological work now in progress. Special attention is centred on the Centrosaurus bone bed.

An explanation of the park's unique geological aspects (supported by accompanying photographs) prepares the reader for the history and natural history of Dinosaur Provincial Park.

The final section deals with the present set-up of Dinosaur Park. It should be noted here that most of the park is not open to the general public. This is not meant as a deterrent, but enables proper supervision and protection of the valuable resources found in every nook and cranny.

So do not expect to see the ruins of the Vision Quest Site or Happy Jack Jackson's homestead. Only specially guided tours or people with government permits have actually viewed them. Perhaps some day suitable access to these sites will be established. However, Dinosaur Provincial Park is expansive enough to satisfy the average adventurer.

It was not my entention to be overly scientific in my approach to this book. There are many scholarly books available on geology and related fields (see Bibliography).

I hope *Dinosaur Provincial Park* will serve as a souvenir and spark a further interest in exploring the mysteries of dinosaurs, our country's wonderful hidden treasures and our rich national heritage.

*Gordon D. Reid    Oakville, Ontario - 1994*

*The slim look in hoodoos. Notice the delicately balanced caprock.*

*Seventy to Eighty million years ago the badlands of Alberta closely resembled today's Florida Everglades.*

# Part One

## The Dinosaur

Imagine a low, marshy area consisting of small subtropical lakes, streams and deltas. That was how Dinosaur Provincial Park appeared during the Upper Cretaceous period (Mesozoic Era), 70-80 million years ago, on the northern extremity of the once-great Bearpaw Sea. The Bearpaw extended northwest from the Gulf of Mexico, covering most of south-eastern Alberta, southern Saskatchewan, southwest Manitoba and large sections of the United States.

Dinosaur Provincial Park was a veritable tropical paradise, but it had its share of chaotic moments in natural history. Floods were frequent. Volcanic ash from as far away as southern British Columbia and Washington State blanketed the ground. Torrential downpours followed. The earth was still in a tremendous period of transition.

The humid climate, somewhat like the Louisiana Bayou country of today, supported lush vegetation. Swamps and marshes abounded with cattails, which camouflaged the primordial slitherings of turtles, crocodiles, rays, salamanders, sharks, sturgeons and small fish-eating reptiles. This prehistoric everglade also supported the long-necked Plesiosaurs and Mosasaurs, which grew to lengths of ten metres (33 feet).

Inland, the dryer grounds made way for giant Metasequoia (ancestors of the redwood), tree ferns, breadfruit, sycamores, chestnut, magnolias, fig trees and other ubiquitous flora.

Today Dinosaur Provincial Park boasts over 30 different species of dinosaurs. Creatures like Albertosaurus, Stenonychosaurus and Lambeosaurus once foraged and fought together here, striving to co-exist in a wild and primitive land. These giants of the past existed for ten million years before their mysterious demise. Over 500 museum-calibre specimens have been removed from the Red Deer River valley and are on display at institutions the world over.

Of the four different dinosaur levels in Alberta, the Oldman formation at Dinosaur Park has provided paleontologists (also known as paleobiologists) with an abundance of fantastic finds.

*Ammonite molluscs (large shell in photo) were a major source of food for Mosasaurs in the late Cretaceous period.*

*Albertosaurus (mid ground) hunts near stream's edge as a startled Champsosaurus (foreground) seeks the safety of the water.*
*~ National Museums of Canada*

"At one time, paleontologists were more concerned with what these animals looked like", said Dr. Philip Currie in *Canadian Geographic* magazine. "Now the emphasis is placed on how they lived. Were they warm-blooded like mammals? Did they care for their young? Were the mighty carnivores hunters or scavengers?"

Dinosaur Provincial Park continues to amaze Dr. Currie of the Tyrrell Museum, Drumheller: "We've been averaging probably one new type of fossil animal in Dinosaur Park for the last ten years. This year (1985) we blew it all away!

"We've been looking pretty hard at our collections. We have come up with at least four new species this year for Dinosaur Park. They are pieces, not full skeletons. Two of the bone-headed Pachycephalosaur (thick-headed lizard) dinosaurs and two types of dinosaurs from Mongolia. Erlikosaurus (Erlik's lizard) had a beaked jaw with short, sharp teeth – which is not a type of dinosaur we had found in North America at all. It is a prosauropod, which are common in Europe during the Triassic period. They show up in the late Cretaceous in Mongolia. Another carnivorous dinosaur, called Elmisaurus (foot lizard), also came from Mongolia." Elmisaurus was a thin, medium-sized flesh-eater with smaller hands than Dromaeosaurids.

Most of Dr. Currie's recent work has dealt with his fascination for small carnivorous dinosaurs. "There's a lot more diversity there than you would suspect. They are probably warm-blooded. They are intelligent in big brain forms; probably a little more interesting, as they are from a group that is ancestral to birds. It is always more fascinating when the group is not dead end – that it leads to something else."

*Lambeosaurus*

Albertosaurus (Alberta lizard) was formerly known as Gorgosaurus, undoubtedly spawning the 1961 British horror movie *Gorgo*. It was a distant cousin of the dreaded Tyrannosaurus rex (tyrant lizard).

Seven and a half metres (25 feet) in length, Albertosaurus utilized its quick bipedal movement to great advantage, ravaging smaller herbivores with its solid two-ton mass. Horned and armoured dinosaurs were probably looked upon as a delicacy to this fearful monster.

"Very often if you've got two parts of a skeleton from different areas you give them different names. But when you find out they are the same animal, then the older name takes precedence. Albertosaurus (1904) is an older name than Gorgosaurus (1917). Albertosaurus was actually more common than Tyrannosaurus rex in Alberta," claims Dr. Currie.

*Dr. Philip Currie at the Royal Tyrrell Museum of Paleontology ~ Royal Tyrrell Museum*

Daspletosaurus (frightful lizard) stood five metres (16½ feet). This four-ton cousin of Albertosaurus was a bulky, muscular bipedal carnivore. It had small arms like the other large flesh-eaters, but had more teeth than Tyrannosaurus. Daspletosaurus was probably one of the most savage carnivores. Remains found at Dinosaur Park are extremely rare.

Dromaeosaurus (running lizard) was similar to Montana's Deinonychus, but smaller in stature. This 1.8 metre (six-foot) marsh dweller was extremely agile. Its claws held its victim as if in a vice grip. Undoubtedly it was one of the fastest carnivores.

Smaller, swifter flesh-eaters were once numerous at Dinosaur Park. They include Stenonychosaurus (narrow-claw lizard), Dromiceiomimus (emu mimic) and Struthiomimus (ostrich mimic).

Many species of Hadrosaurids (duckbills) have been found at the park. They included Corythosaurus (helmet lizard), Lambeosaurus (Lambe's lizard), Parasaurolophus (similar to Saurolophus) and Kritosaurus (big lizard).

Named for their broad, flattened skulls with flat beaks, the duckbill's jaw contained hundreds of overlapping peg-like teeth. These many grinding teeth were necessary to replace worn ones, as they probably spent most of their time consuming large quantities of vegetation — or swimming away from predators. The hollow crests over the duckbills' skulls suggest they signalled one another with bugle-like warning cries when predators tried to attack the herds.

Centrosaurus (sharp-pointed lizard) is the most familiar genus at the park (see: Excavating the Centrosaurus Bone Bed). Growing up to six metres (20 feet) long, its nose-horn tilted forward. The

*One of the displays you could expect to see when visiting the field station at the Park.*

Centrosaurus lived in herds and it is believed they looked after their young.

Ceratopsians like Chasmosaurus (cleft lizard) and Styracosaurus (spiked lizard) required their horny armour for defensive purposes, probably butting rival males. Cycads and spiky palm fronds may have been their staple diet, judging by their beak-like skulls and shearing teeth.

Ankylosaurs (armoured lizards) like Panoplosaurus (fully armoured lizard) and Euoplocephalus (well-armoured head) were small, squat, slovenly creatures with flexible armoured plates and spikes for protection. The Ankylosaur jaw was weaker than that of the ceratopsian, with much smaller teeth. It is believed they ate soft shrubbery plants and possibly insects.

*Ankylosaurs like this one were extremely wary of predators when feeding. ~ Photographed at the Royal Tyrrell Museum*

Other lesser-known species of dinosaurs have been found at Dinosaur Provincial Park. Recently Dr. Currie's Tyrrell team found a small herbivore purported to be a dinosaurian equivalent of a mountain sheep.

Flying reptiles like the Pterosaurs had very fragile bones. There is no doubt both birds and Pterosaurs once existed at the park. One known flying reptile (from Mexico) was Quetzalcoatlus, which had a wingspan not unlike that of a Second World War fighter plane!

In 1980 Dr. Dale Russell of the Museum of Natural Sciences, Ottawa, visited Dinosaur Provincial Park. That summer he was the first paleontologist to unearth the limb bone of a Pterosaur (estimated wingspan 12 metres or 39 feet), which is believed to possibly be Quetzalcoatlus. These nightmare creatures flourished for about 70 times the present span of mankind. Now they are gone. Scientists and amateurs alike have often theorized about what exactly happened.

Dr. Currie's leanings are towards a natural phenomenon. "Dinosaur Park is at least ten million years before the end of the age of dinosaurs, so there is another ten million years before the theories. I don't think the evidence is all that strong right now for a catastrophe or slow death for dinosaurs. There are a lot of good theories floating around and a lot of good evidence, but it's really in a preliminary phase, like so much of dinosaur paleontology. We require hundreds and hundreds of man hours being dumped into it and it is never going to come."

Other paleontologists believe in the idea of a supernova or an asteroid striking the earth, sending iridium into our atmosphere.

*Chasmosaurus (a cousin of the Centrosaurus) were also thought to roam in herds in what is now Dinosaur Provincial Park.*

More recent theories support death by a plague of diarrhea or earthbound comets. Overall, Dr. Dale Russell commented in *A Vanished World: The Dinosaurs of Western Canada,* "They (dinosaurs) changed as their world changed for 140 million years."

Following a new group of plants and resultant climatic changes, with an annual alteration of warm and cold seasons, the dinosaurs seemed to cope. Then, according to Dr. Loris S. Russell of the Royal Ontario Museum, "The winter lows began to approach the critical level, for the dinosaurs could not hibernate like their cold-blooded relatives and they had no insulation to prevent loss of body heat . . . incubating eggs too were vulnerable . . . a succession of winters, mild by our standards, but critical for dinosaurian survival, might have continued for several thousand years before bringing about total extinction, but in the geological record this is a moment in time."

*Phred, the prehistoric buffalo, is a permanent relic you will encounter on the bus tour. Ask a tour guide about his mishap.*

# Glaciation of Dinosaur Provincial Park

Sixty thousand to 100,000 years ago the Classical Wisconsin glacial ice mass (some 2,000 feet or 610 metres thick) covered most of Western Canada. It reigned supreme for 8,000 to 10,000 years.

The Dinosaur Park area was relatively flat. The exposed land surrounded by short-grass prairie was "gouged" out, as if the claws of some horrendous dinosaur had gashed open the earth's surface, wishing its ancestors to be exposed for future generations. The meltwater channels carved up the land, but the catalyst was the Red Deer River, resulting in the present badlands system.

During the final retreat of the Wisconsin ice sheet, drainage was forced off to the southeast rather than to the east. For a while emphasis was on the streams that flowed in a southerly direction.

Dr. Russell explains, "Then the re-erosion of the valley started. Because there was a valley there in pre-glacial time, it was easy to erode. Semi-arid conditions at times kept vegetation from taking over. Why there is such a rich assemblage at Dinosaur Park is a matter of the situation that developed. There may well be lots of other places buried under the prairie which are just as rich. It's partly a matter of exposure. You will get the fossils if you are in the right formation."

The Wisconsin ice sheet wiped out earlier traces of smaller, ancient glacial systems. The Wisconsin was one of many that traversed Dinosaur Park. It was also the last.

According to Dr. Russell,"Dinosaur Park is peculiar in that everything seems to be about the same age. It hasn't been possible to establish any succession. The actual thickness of the Oldman formation is not very great as compared with the Edmonton group to the north. Lots of specimens are there, but the geological aspect has not been developed. Most interesting are the ecological associations and the conditions under which these animals were killed and buried. This is called 'taphonomy,' the study of how things are buried, a science of funerals."

Deltas and variant rivers and streams at the park changed course so often that the balance of nature was constantly disturbed. These river erosion processes deposited sundry degrees of sediments, spreading the remnants of decayed prehistoric beasts, so that now only fragments of dinosaur skeletons are usually found. Whole skeletons are rarely discovered. Incomplete remains are generally a result of scavengers, who preyed upon the fallen.

One might speculate that the Dinosaur Park area had become a "catch-all" for the dinosaurian remains which eventually settled downstream. This may be true to a degree, but the Dinosaur Park area is more likely a graveyard representing its own unique genera. There the carcasses were caught in eddys, sandbanks and deltas. Sands and silts covered the bodies. The bones absorbed calcium, silica and carbon, beginning the slow process of fossilization or permineralization.

# Excavation Work

The exacting task of properly unearthing articulated remains is carried out meticulously by paleontologists and crew members.

Museum staff and volunteers (some Albertans, but others from various corners of the world) rise at 5:30 a.m. for the short drive to the ever-waiting, ominous badlands. A two-kilometre hike, 20 minutes over rough terrain, awaits the crew as they unload their trucks. Crews vary from six to 20 people. Packs and rucksacks are loaded with tarps, brushes, geologist's hammers, chisels, awls and water canteens in readiness for the hours of toil ahead.

Even though the workers labour in soft sandstone, one might spend an entire day excavating one metre of ground. Frustration may set in when only one or two bones are removed from a few cubic inches of rock. Brushes and dental tools are painstakingly used to reveal minute detail. *Butvar* (glue) is poured on and into brittle bones to preserve them for removal. Butvar also serves as a solidifier.

The volunteers are persistent. Eight-hour days with half-hour lunchbreaks do not deter their undaunted spirits. Flies, mosquitoes and searing heat (sometimes reaching 45° Celsius or 113° Fahrenheit in the shade) do not seem to bother the team. They are proud to be working in one of the three largest bone beds known to man, the Centrosaurus bone bed. (The largest is in East Africa.)

Dr. Philip Currie took an active interest in the Centrosaurus bone bed during the first two years of its excavation. But as the Tyrrell Museum grew, it became more impossible for him to find free time.

*Philip Currie and an assistant paleontologist, Darren Tanke, meticulously clean a Ceratopsian horn core.*

*Volunteers work diligently under a hot midday sun to unearth specimens at the Centrosaurus bone bed.*

*One of many Centrosaurus bone beds, excavated at Dinosaur Provincial Park.*

*Duckbill limb bone eroding on bentonite surface.*

*An unidentified vertebra fragment. Note the dime on the bone to indicate scale.*

Dr. Currie instructs his staff as to exactly what he requires from the bone bed. He checks the results every two or three weeks.

Volunteers work three-week stints at the park. It takes time to train their eyes and teach them to utilize tools properly. Anyone interested in becoming an unpaid volunteer may write Tyrrell Museum of Paleontology, P.O. Box 7500, Drumheller, Alberta T0J 0Y0. No related training or experience is necessary.

Dr. Currie comments, "The volunteers come from different backgrounds. Sometimes they come up with very perceptive ideas; some work for museums and some are students of paleontology. They are always learning something. The volunteer program is wonderful. It's one reason why we get as far as we do."

Volunteers and staff systematically map out bone associations using an archaeological grid system with string or rope, which is staked out in ten-centimetre (four-inch) squares, overlaid to provide points of reference.

Once bones are revealed, a "metre box" is placed over the area. The metre box is a wooden frame measuring one metre (39.37 inches) square. It is marked off in tenths. A base line is used to guide the box for fossil orientation. Fossils on each square are graphed onto a map one layer at a time. Each layer represents a depth of 15 to 20 centimetres (six to eight inches). The bone bed has required several maps to keep consistent cataloging in order. Prior to completion of a sector, the crew covers their work with plastic sheets to protect the fossils from erosion.

*The precision of a helicopter lift has replaced the strain and delays of the old horse and wagon transporting method for fossilized remains.*

Removal of bone specimens is arduous yet important work in the museum's collection system. Paleontologists and sedimentologists number the specimens, cataloging them right on the field. There are three park references: the year, the locality and the batch (specimen) number. The first two numbers go on the plaster- and burlap-coated jackets; the specimen number is put on the fossil before jacketing, as well as on the outside.

When the bone bed nears completion Dr. Currie hopes a 100-square-metre (1,076-square-yard) sector of exposed bone will be housed for public display. Most bones are removed as soon as possible upon exposure to the air, otherwise erosion would destroy specimens within a month.

Decades ago, Lambe, Brown and the Sternbergs used horses, dynamite and boats for efficient removal of fossil finds. Dr. Currie remarks, "We've got jack-hammers, pneumatic drills and air compressors, but we still can't get a compressor around the badlands. When it comes right down to it, the most useful tools are still chisels and hammer. We've been lucky because we've had helicopter money to get specimens out. It is really efficient. We can take out as much as ten times the material or more in a couple of hours. The cost per hour may be high, but it works out a lot cheaper compared to the 'old boys' and their hired labour system.

"We don't just excavate a skeleton (isolated ones) and remove it right away. We have to wait until we've got enough specimens to justify the expense. Whereas the old boys pulled them out when they finished. Their horses were good for access too, because essentially they could go anywhere in the badlands.

"Dynamite was very effective – removing a lot of rock very quickly. A jackhammer is a lot slower. It's more physical too."

Helicopter work means a well co-ordinated plan. Usually only one helicopter is available. Manpower is limited, so people must be transported from one specimen site to the next in unison with the chopper's pre-organized movements. Specimens are then removed in sequential order.

"It's exhausting work," says Dr. Currie, "taking three to four hours to get the specimens out. Some plaster jackets weigh 600 pounds to a ton. The helicopters never land. They hover over us until we hook the net onto the chopper and get the hell out of there. Then he lifts as soon as he sees us in the clear. While he's gone we load the next net."

The large commercial helicopter transports its precious cargo some nine kilometres (over five miles) across Dinosaur Park to the lookout area and the waiting flatbed trucks. People at the trucks move the nets, which then emptied are placed back on the helicopter. Cherry picker units on trucks turn and manoeuvre jackets into proper allocated spaces. The nets are taken back and dropped on the next site. This operation requires a constant shuf-fling, back-and-forth process, moving people and supplies from one quarry to the next.

# Excavating the Centrosaurus Bone Bed

*A Centrosaurus, at the Calgary Zoo's Prehistoric Park, calmly feeding on plants.*

Although Charles H. Sternberg, Barnum Brown and other noted paleontologists probably discovered the now-famed Centrosaurus bone bed at (Quarry 143) Dinosaur Provincial Park, they overlooked it simply because the removal of whole articulate (clearly jointed) skeletons was then in vogue. The Centrosaurus bone bed is significant in that it is the first to be excavated to such an extent at the park.

In 1977 two park naturalists were out on a hike, foraging for a variety of things, including dinosaur bones, when they discovered a few bone fragments that had weathered out onto the surface. At the centre of this 2,500-square-metre (26,880-square-foot) bone bed, the exposed bones stretched out so thick that there was no exposed ground. On further inspection it was obvious that six animals had died on the spot, based on the varied skull and bone sizes.

"The really interesting thing was the number of cerotopsian dinosaurs, which is disproportionate to what you would expect. It looked promising. A good place to start working on a bone bed in Alberta," says Dr. Currie.

Work began in July 1979, but by early 1981 a minimum count of 38 individuals had been ascertained due to the amount of partial skulls discovered. Seventy-eight square metres (840 square feet) had been excavated. The year before, 50 other bone beds had been documented at the park.

To date, over 55 animals have been found. Scientists the world over have come to seek clues, wondering what actually brought about the demise of such a herd. The bed is one of five in the park

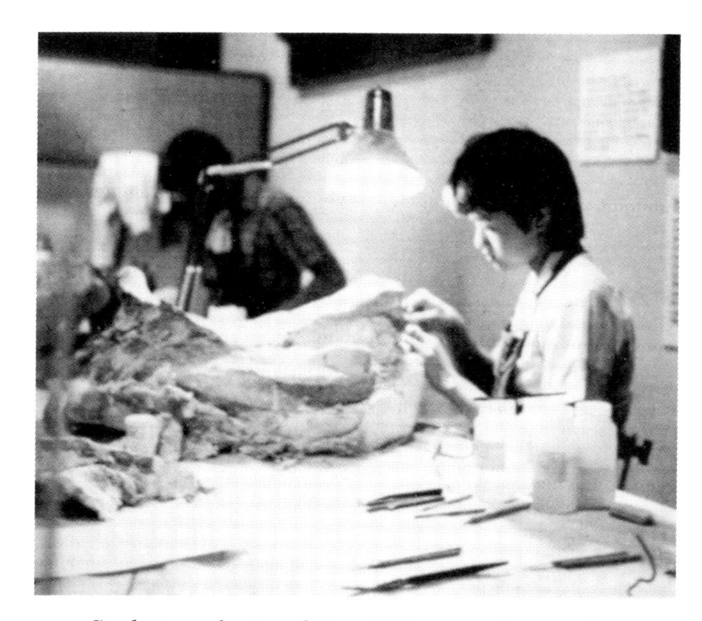

*Students at the Royal Tyrrell Museum, Drumheller, preparing Centrosaurus skulls for future exhibits. Note Butvar bottles at right (used to glue brittle portions of bone).*

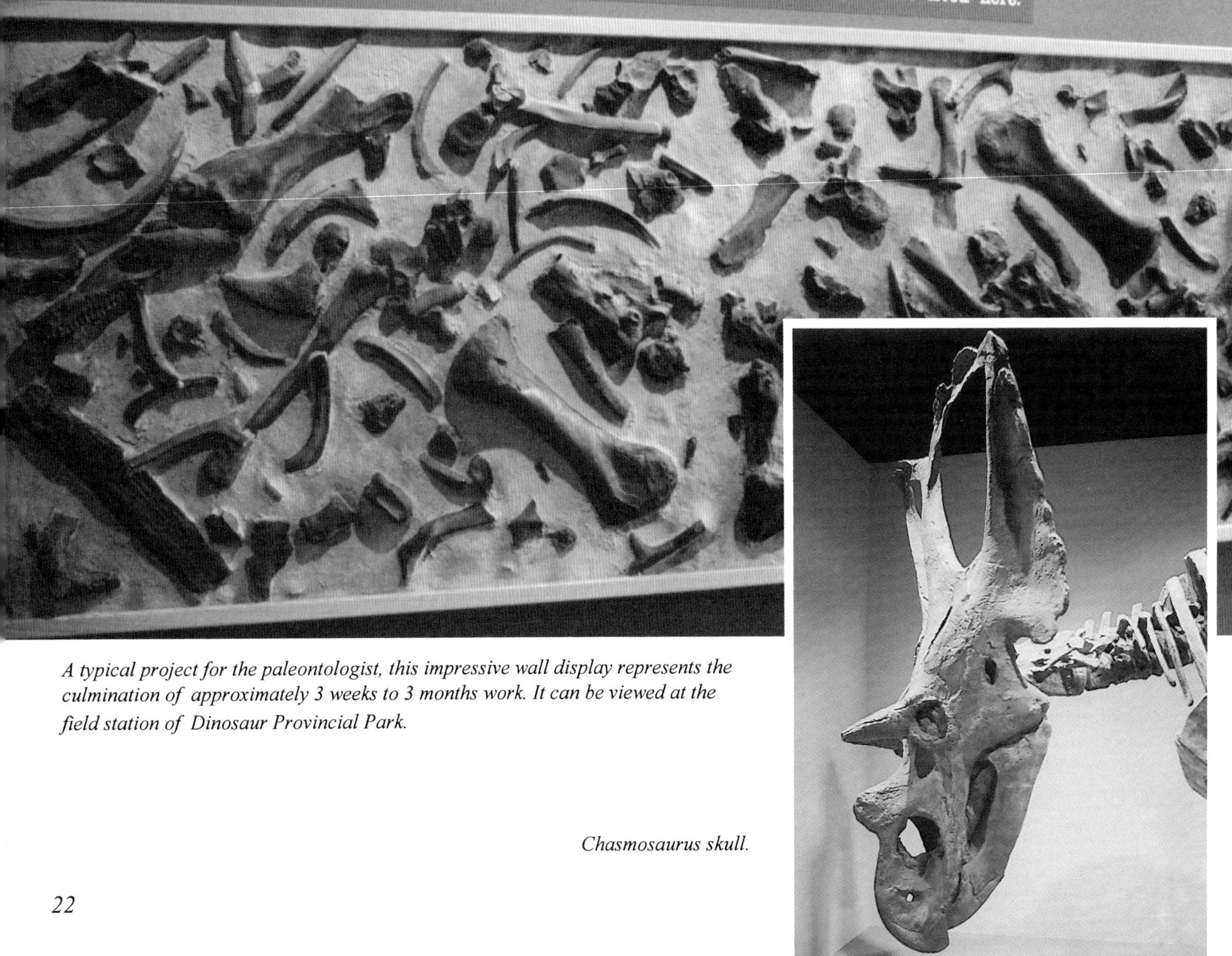


*A typical project for the paleontologist, this impressive wall display represents the culmination of approximately 3 weeks to 3 months work. It can be viewed at the field station of Dinosaur Provincial Park.*

*Chasmosaurus skull.*

*Full Champsosaurus skeleton as portrayed in the painting on page 10.*

*The terrible dragon sleeps. Seventy million years ago this Albertosaurus stalked the forests and deltas of Dinosaur Provincial Park. What is left in stone reveals to the observer that this creature was a tyrannosaurid, one of the most formidable of all the predatory dinosaurs.*

with ceratopsians. One of the beds, about a kilometre away, is twice as large.

"You don't get full specimens in that bone bed," Dr. Currie points out. "Everything is disarticulated and mixed up, jumbled. You may find all the bones of a skeleton, but you don't know which individuals they belong to. We take bones that are unique to the body – something like the horn over the nose. Actually what we've used is the right orbital horn. The horn over the eye. Of course every animal has only one right eye!"

As many as 60 bones and fragments per square metre have been yielded. It is one of nine Alberta bone beds exposing a 90% generic similarity. Up to 400 specimens may have come to rest there.

What actually happened to this herd of Centrosaurs is a matter of conjecture. It is believed the ceratopsians were victims of a river flood, much like the 3,000 caribou that drowned en masse a few years ago in Quebec.

Bone beds, which consist of bones of other species mixed with the main animal (Centrosaurus), can be formed in different ways. Most of the bone beds at Dinosaur Park are long-term accumulations that may represent thousands of years of bones piling up on one another and eventually washing into a single river plane or dumping place.

During the seasonal variation (evidenced by growth rings in tree fossils) at the park, there was some sort of flood phase, with high river levels, resulting in specific layers of sand over a year. All the bones in the river were laid down over a short period of time.

Dr. Currie's suspicions are that "the sequence was that the animals tried to cross the river while migrating. Dinosaurs almost certainly herded. They were subject to natural catastrophes. As individual animals they could probably swim quite well, but when you get a lot of animals jostling with the others, then they're going to drag each other down."

The carcasses drifted downstream until stranded on sandbars or the river's edge. The river waters lowered as the flood ended. Then the big carnivores like Albertosaurus moved in, tearing and clawing cadavers apart. Evidence of the scavengers is apparent from samples of teeth, broken bones, spiral fractures on bones and tooth-marked bones found at the site.

Exposed to the sun during the ensuing dry season, the skeletal remains rotted away. The following year the river probably entered a flood basin, picking up all the bone, slightly moving the larger ones, towing them the same direction as the current, more or less perpendicular in a northwest-southeast direction. All the smaller pieces – isolated teeth, little toe bones, tail ends and other bits – would wash downstream, right out of the system.

According to Dr. Currie, "You get a dinosaur that looks different from another one, the tendency is to call it a different species. But it could just be a male or female of the same thing, versus being babies and adults. You've got to have that repetition. If you've got repetition you can solve a lot of problems. No specimen is ever preserved the same anyway. This is a valuable resource. We've (scientists) got a problem in that we have to priorize what is most valuable, because there is no way to ever do all the work that has got to be done."

Actual sizes of skeletons range anywhere from 1.5 to 5.5 metres (five to 18 feet). Disease has not been ruled out as another hypothesis for the mass death of the Centrosaurs.

# Part Two

## Geology of Dinosaur Provincial Park

One can only appreciate the natural beauty of Dinosaur Provincial Park by understanding its delicate geological balance.

Driving many miles over the great expanse of Alberta's plains, valleys, and grasslands does not fully prepare the new visitor for the sudden drop into "the valley of the dinosaurs." The scenery changes drastically to the snake-like Red Deer River valley curving in amidst a barren lowland area.

Dr. Loris S. Russell's impressions of Dinosaur Park are vivid: "The first time I saw the Steveville area was in 1928. It still continues to astonish me; such an abrupt change. I'm sure everyone gets the feeling of surprise as you come over the rim."

The sedimentary rock at the park is 100 times softer than the majestic Rockies to the west. Erosion at the park occurs ten times faster than the Rockies. Seventy million years ago the Rocky Mountains were just breaking through the earth's crust.

The park entrance, bordered by two scenic viewpoints, welcomes the visitor to a bizarre aerial vista. The Red Deer River valley is best known for its bedrock cliffs, offering a stimulating visual expanse as it permeates 120 metres (400 feet) down below prairie level. The actual canyon cuts about 1.6 kilometres (1 mile) across in a north-to-southward direction. The Red Deer River stretches 24 kilometres (15 miles) through the park, splitting the badlands.

In an estimated 10,000 years this meltwater

*Dr. Loris S. Russell at the Royal Ontario Museum.*

*Aerial view of Dinosaur Provincial Park.*

*The natural processes of weathering denudes the badlands, creating an eerie, otherworldly appearance.*
*~ René Barendregt*

channel from the Wisconsin Glacier will suffer extensive wind, water and frost erosion, wearing away the bowl of the badlands until nothing is left. The canyon's north-southward erosional process tears at the surrounding prairie, working methodically, subtly like the hands on a clock.

The badlands of Dinosaur Provincial Park derive their name from the French *maúvaises terres à traverser*, coined by the French traders who travelled through the Dakotas and Montana, encountering country that was so rugged and so worthless (for grazing and farming) that they called it "bad lands."

Three important creeks empty into the Red Deer River from the park environs. They are the Little Sandhill Creek (it flows through the picnic and campground areas); One Tree Creek, which flows through privately owned land, including badlands; and 4.8 kilometres (three miles) west, upstream and on the north side of the Red Deer River, a half kilometre upstream from Steveville, is Berry Creek.

A popular badland sight is the hoodoo. Described originally by Europeans as "strangely sculptured rocks," the name is derived from the African word "voodoo." These awesome pillars of sandstone are revered by the Blackfoot and Cree Indians, who believed hoodoos were petrified giants who came to life at dusk in order to hurl rocks at intruders.

Hoodoos range from thimble-size to gigantic proportions. They are created by rain erosion along valley and gulley walls. The sediment ultimately wears down to a point where a protective caprock is formed on top of the structure. This is caused by a hard piece of sandstone (fossil bone or pebbles), which retards the erosion process. The softer rock below erodes all around, resulting in these free-form pillars. In general, hoodoos are seen in various

*One of the interesting cut away rock formations. A fine example of erosion in the badlands.*

*An excellent example of strange hoodoo erosional remnant in the Valley of the Dead Lodges.* ~ C. Wallis

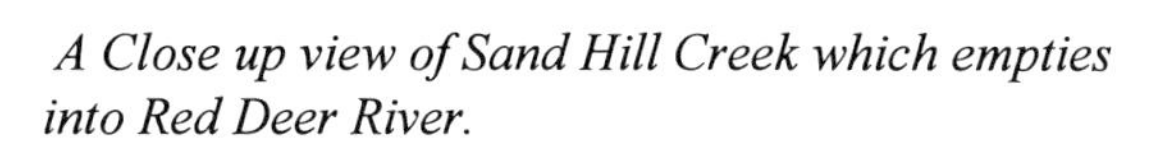

*A Close up view of Sand Hill Creek which empties into Red Deer River.*

forms of disintegration. When the caprock (capstones) fall from atop the pillars, the unsheltered hoodoo erodes quickly.

The actual park rainfall is only 20 - 23 centimetres (eight or nine inches); combined with snow, it totals an average 30.5 - 35.5 centimetres (12 - 14 inches) of precipitation per year. A typical rainstorm could dump anywhere from .6 to 1.3 centimetres (one-quarter to one-half inch) of rain in about 15 minutes. Torrential downpours or thunderstorms moving through the park in 20-minute sweeps often make walking or driving along the badland routes almost impossible.

The bentonitic clay (originally from the weathering process of volcanic ash and used as an oil well lubricant) actually absorbs water to such a degree that it swells to twice its volume, making everything extremely slippery. Park hikes and bus tours are cancelled if the rain persists. The ensuing drying process does not take long. Then the bentonite hardens to a "popcorn-like" surface. Prior to that, wet particles approximately one metre beneath the ground can still inflict startling results on the unassuming hiker!

The badlands rate of deterioration is very rapid. This state of erosion occurs primarily during spring runoff, causing tiny rills to form on the surface of hills and valleys. Given time the rills will enlarge to create countless gullies, sugar loafs, ravines, coulees, jagged divides and ultimately canyons. Surface rain water, known as sheetwash, carries clay and sediment debris away rapidly. Evidence of dendritic (branching) channels or rillwash are seen all over the park. This is why vegetation is almost totally absent, as showers wash away material before plants can gain a foothold.

"Piping" (pipe-ing) is a geomorphic agent dominant in sculpting the badland terrain. It refers to the subsurface removal of sand and silt particles via the action of running water, which leaves behind a network of interconnecting underground voids and tunnels that collapse with time. This process, indigenous to arid and semi-arid regions, is responsible for major gullies being formed in all the badlands of southern Alberta.

"Sink holes" are a result of gulley heads formed by widened pipings which collapse to form crater-like surfaces. They look like caves to the untrained eye.

Pipes vary in size and depth, with diameters exceeding several metres. Deep piping systems come from abandoned upper systems. These systems occur along valley walls, from prairie level above, down to the terraces in the valley bottom below.

Pipe systems will eventually destroy themselves by excessive enlargement and complete collapse. Therefore, it is no wonder that park staff frown on visitors wandering too far into unknown areas of the badlands, such as the Natural Preserve, which is restricted to guided tours only.

"To the casual observer, the processes of weathering, denudation, scarp retreat, glacis formation and piping may not be immediately evident, but rather the other worldly appearance of the badlands may offer the greatest appeal," notes Dr. René Barendregt of the University of Lethbridge. "Combined with the rich supply of fossil material contained in the bedrock, the beautifully sculptured badlands offer a plethora of ever-changing forms and moods to both visitor and scientist. Even the most resistant of formations soon erode or weather to oblivion."

# Part Three

## *History of Dinosaur Provincial Park*

Long after the dinosaurs had vanished, thundering herds of buffalo roamed the open prairie near what is now known as Dinosaur Provincial Park. With the buffalo came the hunters. Roughly 8,000 years ago prehistoric man dwelled in the area. His existence is a mystéry, but he was undoubtedly a distant relative of the Cree and Blackfoot nations.

At its height, the Blackfoot nation (confederation) extended from the Rocky Mountains in the west, over to the Saskatchewan border and north to Red Deer. The Blackfoot Peigan tribe went as far north as Edmonton. Tribes were small, with approximately 2,500 natives maintaining the land of the northern Plains Indians.

The Blackfoot nation was north of the Red Deer River. Dinosaur Park was known by the Blackfoot as "valley of the dead lodges" (Dead Lodge Canyon). This odious name hailed from the aftermath of a Peigan assault on the Shoshoni (Snakes), who lived to the south.

The Shoshoni originated from the United States foothills. They were constantly at war with the Peigan tribe and the Blackfoot nation. The park was home for Shoshoni lodges.

Saukampee, a Peigan chief (so the legend goes), sent scouts to reconnoitre the Snakes camp. There was no sign of sentries on guard, so they decided this was the right time to attack! Saukampee's story appears in Fraser Symington's *The Canadian Indian* as follows: "At the dawn of day, we attacked the tents, and with our daggers and knives, cut through the tents and entered for the fight; but our war whoop instantly stopt, our eyes were appalled with terror; there was no one to fight with but the dead and the dying, each a mass of corruption. We did not touch them, but left the tents, and held a council on what was to be done. We all thought the Bad Spirit had made himself master of the camp and destroyed them. It was agreed to take some of the best of the tents, and any other plunder that was clean and good, which we did and also took away the few horses they had, and returned to our camp.

"The second day after this dreadful disease broke out in our camp, and spread from one tent to another as if the Bad Spirit carried it. We had no belief that one man could give it to another, any more than a wounded man could give his wound to another. We did not suffer so much as those that were near the river, into which they rushed and died, but in some of the other camps there were tents in which every one died."

The Peigans believed a Bad Spirit had entered their camp. Needless to say, the spirit of the Blackfoot nation was broken by smallpox. Some of the old cottonwood trees lining the Red Deer River were used as "burial grounds" for the nation. Platforms were built. The dead were set aloft on the cottonwood limbs. This practice carried on until about 1870 or 1880, yet nothing remains of this ritual. No trace.

*This 6.7 metre (22 feet) long "Effigy Man" could be of religious significance but is now shrouded in mystery.* ~ *C. Wallis*

*Plains indians inhabited locals near the badlands as is evidenced by the stone relics shown on these two pages. Blackfoot brave (reference for costume from a painting by George Catlin, Smithsonian Institution).*

*Vision Quest site (one metre across) was of great spiritual*
*ue to the Blackfoot tribes.*

*An aerial view of Sand Hill Crees as seen from one of the park's more easterly hiking trails.*

The natives respected the valley of the dead lodges, but they did not fear the area. The Blackfoot name for the badlands was "hills on hills." Evidence of their reverence for the area is apparent in the hand-made stone structure of an "Effigy Man." Stretched out over the prairie, it is 6.7 metres (22 feet) long. Shrouded in the uncertainty of time, "Effigy Man" could have been of astronomical or religious significance, but no one is certain.

Equally fascinating are the remains of approximately 500 tipi rings (stones used to weight down the edges of tipis), generally found along the high plains overlooking the river. These circular remnants mark generations of various tribal inhabitants. The tipi rings range in size from six metres in diameter to one or two metres. The smaller rings were probably sweat lodges.

Sweat lodges consisted of willow branches stacked and interwoven with buffalo robes. Hot steaming rocks began a ritual involving a brave with his medicine bag. A spirit from a "Vision Quest Site" would instruct the brave how to select items for his medicine bag. The items could supposedly conjure up power for the brave, giving him a sense of oneness with his spirit world.

To the untrained eye a "Vision Quest Site" is just a circular pile of rocks (about one metre across) atop a hill. To the natives the site was a symbol of religious stability.

Vision Quest Sites were generally located at the highest point possible. Water had to be to the north (if none was available, then to the south). The site had to have a clear, unobstructed view of the horizon in order to see the sunrise or sunset. It is not certain who originated the site, but the Blackfoot, Cree and Shoshoni probably all shared it.

Park naturalists have found a pipe at the site. This artifact more or less proves its religious value. Young teenage braves visited the spot with a little piece of leather to sit on and a blanket, but little else. The brave had to stay there without any food or water, weathering the summer heat. The boy would wait from two to five days until he saw a vision. Some braves never lived to reach manhood.

In 1984 Dinosaur Provincial Park had a record high temperature of 51° Celsius (124° Fahrenheit). To stay outside for even a short while at that temperature, without movement, food or drink, the bugs eating you, would assuredly cause some kind of vision!

Vision Quest Sites tend to point towards something important. The one in the park seems to point towards "Effigy Man."

The only other Indian relics unearthed at the park were a spearhead and two arrowheads found inland near "the Citadel," a park formation. There is no record of the Indians venturing further into the park. The natives dwelled on the plains, only entering the valley to hunt, bury their dead or seek their visions.

# The Dinosaur Hunters

"The wisdom of man is foolishness to God."
— Charles H. Sternberg, 1917

On June 9, 1884, Joseph B. Tyrrell explored the Red Deer River valley, methodically advancing to the Dinosaur Park area. All he had was a wagon and a canoe. "When he looked up the valleyside and saw a great ugly face with rows of 'sharp, spikelike teeth' grinning down at him he knew for certain that nothing like this had ever been found before," wrote Alex Inglis in his *Northern Vagabond: Life & Career of J.B. Tyrrell.* Tyrrell, who now has a paleontological museum named after him, was working for the Geological Survey of Canada (GSC). His "ugly face" discovery turned out to be an Albertosaurus!

Thomas C. Weston, another GSC geologist, followed suit in 1888. Weston worked from a boat, going ashore to search the badlands. But it was not until 1897-98 that Lawrence M. Lambe led two separate expeditions for the GSC.

According to Dr. Loris S. Russell of the Royal Ontario Museum, "People like Tyrrell, Weston and Lambe had no practical experience in collecting fossils. In fact very few people did, even in the United States. But the professional collectors like the Sternbergs and Barnum Brown had been developing techniques. Williston and Hatcher were both professional collectors. They worked out methods of taking up these things in blocks, one way or another, so that the association of the bone fragments was preserved.

"When (Barnum) Brown came in (1910), he brought these techniques with him and also people who had been working with him for years and understood how to do it. This changed the picture completely. Then the Canadian government had to take some action. There was no one in Canada who could handle that problem, so the GSC hired the whole Sternberg family — Charles H. Sternberg and his three sons, Charlie M., George and Levi."

The Survey's vertebrate paleontologist, Lawrence Lambe, was Canada's first scientist to attempt to describe new types of dinosaurs in scientific terms. Lambe toiled well into the early 1900s in the Steveville area of the park. Lambe's detailed studies were developed in co-operation with H.F. Osborne of the American Museum of Natural History in New York City.

Lambe did not see eye to eye with Sternberg Sr. on future work. In 1916 Sternberg left the GSC with his youngest son, Levi. George, the eldest, left in 1918. In 1919 Lambe died, leaving Charlie Sternberg Jr. to maintain operations under Dr. E.M. Kindle, the GSC chief paleontologist.

The "Great Canadian Dinosaur Rush" was heralded in by the pioneer work of Barnum Brown in 1910, even though the GSC had known of the richness of Alberta's fossil fields for over 26 years.

By 1914 the Sternbergs had set up camp in the Red Deer River valley. They had arrived earlier, further north, near Drumheller in 1912. Their "hunting grounds" were set up about eight kilometres

*All that remains of the old Mexico Ranch, once known as the Homestead of Happy Jack Jackson.* ~ C. Wallis

(five miles) east of Barnum Brown's site on the Little Sandhill Creek. Park campgrounds are located there now.

By 1916, despite First World War restrictions, the Sternberg family had systematically excavated and shipped no less than 16 dinosaur skeletons to Ottawa via the old CPR lines (now gone) at Patricia. In fact, the old gravel road which winds down from the twin lookout points into Dinosaur Park was one of the main routes where these dinosaur bones were actually carted out.

Sternberg Sr. continued work for the British Museum with his son Levi. In 1916 two of their duckbill specimens were lost when a cargo ship, the *Mount Temple*, carrying the precious bones, was torpedoed and sunk by a German submarine.

In 1917, the final year of the "Dinosaur Rush," Sternberg Sr. wrote his memoirs of the Red Deer River days (updated in 1932) extolling the wonders of being a dinosaur hunter in Alberta.

Arriving at their Steveville "camp" (west end of Dinosaur Park), Sternberg wrote, "At ten minutes of five in the afternoon the naked buttes, towers and ridges of the Belly River Series of the Cretaceous loomed up in the distance. We soon reached Steveville and managed to make a landing in the swift stream, just below the Ferry, and below the mouth of Berry Creek on whose border the little town stood. A hospitable town it proved to us; especially have we often enjoyed the hospitality of Steve Hall's Hotel; after this jolly good fellow the town gets its name."

Steveville, with a population of approximately 100, was essentially a wheat town, with the prospects of oil-producing capabilities. The ferry was built in 1911, operating until 1972, when a proper

*Early display, circa 1921, of a Dinosaur's leg bone found below Steveville, Alberta.* *~ Glenbow Archives*

*Dr. Barnum Brown poses beside the articulated remains of a corythosaurus excavated near Steveville, Alberta.*
*~ Glenbow Archives*

bridge was ultimately erected. Steve Hall and his wife Edith developed the site, constructing a boarding house, stables, general store, post office and lumber yard – all essential to the development of a small western community.

Steveville is now a ghost town. The CNR line came south from Hanna to establish a Steveville station. When the railway pulled out the tracks, the town died.

Later Sternberg wrote, "Suddenly, I stumbled on a couple of orbital horn-cores of a new genus of these strange creatures. The nasals and much of the face had been disintegrated by exposure to rain and frost; one complete lower jaw and part of the other was in place, however. With eager hands I used my little pick and digger, cutting into the face of the cliff. The horn-cores were pointed heavenward."

At "Happy Jack Ferry," 20 kilometres (12 miles) east of Steveville, Sternberg had found a "strange spiked dinosaur," Styracosaurus.

Again, names in the area were derived from ranching "legends." Happy Jack Jackson, supposedly a relative of the ill-fated Confederate General Stonewall Jackson (of the U.S. Civil War), was seldom sober. His original homestead covered 65,000 acres. Now it is a shambles of old sod cabins, stables, a workshop and broken-down fences. The badlands, like a decaying enigma, stretch out behind the ruins. The land was originally settled in 1903 by Lord Delaval James de la Poer Beresford and known as the Mexico Ranch. Old Hansel Gordon (Happy Jack Jackson), Beresford's foreman, inherited the sprawling ranch in 1906.

Happy Jack Jackson had travelled northward from Mexico with Beresford on a cattle drive. Jackson continued to run the ranch, the ferry and still maintained an interest in drinking considerably right up until his death in 1942. He was buried at Brooks. Jackson's diary in 1913 was succinct enough:

| | | |
|---|---|---|
| April 23 | Wed. | Drunk |
| 24 | Thurs. | Red Had a Calf 8 Days over |
| 25 | Fri. | Still drunk |
| 26 | Sat. | Sick |
| 27 | Sun. | Dam sick |
| 28 | Mon. | Worse |
| 29 | Tues. | Very Feeble |
| 30 | Wed. | Long Live Booze/Hurrah for Hell |

As a result of local ranching, wandering cattle were a hindrance to paleontologists at times. The rattlesnakes, heat and mosquitoes were hardly noticed next to stray cattle.

The Prince Kerr Ranch (P.K. Ranch) was the first one in the area, established in 1889. Another prominent ranch was run by Black cowboy John Ware, son of a slave. Ware's cabin was later moved 20 kilometres (12 miles) to the park. Ware died in 1905 as a result of an accident with his horse.

In 1935 Charlie Sternberg collaborated with a topographer of the GSC to map the Steveville-Deadlodge Canyon badlands. Actual sites of scientifically described specimens were marked for posterity with concrete-based brass rods. Charlie's son, Raymond Sternberg, accompanied his father in 1936 to finish the job. It meant a permanent record of where articulate remains were found.

Charlie Sternberg Jr. was outgoing. He liked to tell people about his excavations. Dr. Loris S. Russell recalls, "On one occasion, when we were sitting down to a Sunday dinner, a car came up the valley

George Sternberg relaxing near one of his many digs.

*Paleontologist George Sternberg and assistant Kelly excavating dinosaur remains at Steveville Badlands, circa 1921. The chisels and brushes these two men are using, are still considered the most useful tools today!* ~ *Glenbow Archives*

from the ferry and headed past our camp to the mouth of a coulee, where the dinosaur skeleton was.

"We jumped into our Model T-Ford and took off after them. By the time we got there they were just at the point of taking the canvas off the skeleton. Charlie was very good. He was friendly and explained to them that this could cause damage and so on. They were very apologetic. They hadn't meant any harm, just too interested. That's the only occasion that I ever saw local people being a problem."

Charlie M. Sternberg led 13 expeditions to eastern and western Canada before his retirement in 1950. He continued to assist researchers out of his National Museum (Ottawa) office until 1970.

*Above - Deep rill patterns offset by the ever dominant, hauntin hoodoo structures.* ~ *C. Walli*

*Left - Prairie entrance to the park. Gateway to the badlands.*

# Development of Dinosaur Provincial Park

Local ranches were important as landmarks for field parties and undoubtedly paleontologists paid social visits to farmers on occasion.

Local citizens like Dr. W.G. Anderson of neighbouring Wardlow were in favour of Steveville-Deadlodge Canyon becoming a national park, but concern was expressed over the possibility of indiscriminate removal of dinosaur fossil treasures.

Boards of trade from local towns and hundreds of locals put forward a movement to save nature's marvels at the park. And in 1937 the most ardent advocate of initiating a park, Dr. Anderson, organized a pilgrimage. Hundreds of southern Albertans visited the park area. Sand Hill Creek was the main attraction. Levi Sternberg was excavating a complete skeleton there. Heavy rain did not hinder the crowd's enthusiasm. Dignitaries from Edmonton and Calgary arrived, including Tom Baines, who was then curator of the Calgary zoo and their Steveville exhibits.

Dr. Anderson headed a tour of the park area. He related the history of Alberta's dinosaur hunters and was openly critical of the government's indifference towards park development.

Other than protection via the Provincial Parks and Protected Areas Act in 1930, and various investigations throughout the 1930s, the park was relatively open to anyone.

Museum budget cuts and the Great Depression restricted proper scientific exploration. The Second World War set national interest in dinosaurs to an all-time low.

*The unveiling of the above cairn heralded a new beginning for Dinosaur Provincial Park.*

*A legacy for all....*

***UNESCO's** emblem, symbolizes the interdependence of cultural and natural properties. The central square is a form created by man and the circle represents nature, the two are intimately linked. The emblem is round like the world but at the same time is a symbol of protection.*

***UNESCO** (United Nations Educational Scientific and Cultural Organization) is a branch of the UN, committed to the protection of earth's richly diverse cultural and natural heritage sites. **UNESCO** operates on the fundamentally new idea that such protection is not the separate responsibility of each individual country or state but the joint responsibility of all humanity.*

*Dinosaur Provincial Park was officially declared a **UNESCO World Heritage Site** on June 19, 1980. To date there are more than 288 **World Heritage Sites** all over our globe.*

"Emphasis was on the practical exploitation of natural resources," Dr. Russell recalls. "This institution (the Royal Ontario Museum) had very little money for field work. I had plenty of money for my own particular expeditions, which were small-time really, not aimed at collecting large, expensive specimens."

The media co-operated whenever possible. A 1954 Lethbridge *Herald* editorial stated, "For years representations have been made to get a park established in the badlands. It would serve two purposes. It would provide machinery for preserving this treasure-house of the ages for all Canadians and for all time. And it would provide facilities for the people to see, study and enjoy this caprice of nature . . . So far the province has done nothing except to 'pass the buck' to Ottawa."

A scant $10,000 was set aside by the provincial government in 1955. Following preliminary surveys of the park, officials consulted with Charlie Sternberg and decided to choose the Little Sand Hill Creek-Steveville area instead of Drumheller as an official park.

On June 26, 1955, Steveville (Dinosaur) Provincial Park was officially recognized. But a lot of preparation was in store. It was not until the spring of 1959 that the park was opened to the public.

The late Roy Fowler, who passed away in 1975, was Dinosaur Park's first warden. His expertise went back to his amateur fossil-hunting and archaeological treks of the late 1920s.

Dr. Loris Russell remembers Fowler well, "He was quite a man. I first encountered him in 1928. Fowler was still farming at Aldersyde, north of High River, had a shed full of fossils and was very generous with them.

"Later, when Roy had the opportunity to move to the park, I strongly recommended him as the first custodian. He was enthusiastic, without a formal training, but was self taught."

Charles Sternberg and Roy Fowler worked closely together, arranging bone specimen "bird cage" shelter installations. Fowler found a skeleton in the summer of 1955. It was moved, due to its precarious height, and is still on display. Sternberg then found another display-quality specimen, and by 1958 the skeletons were both housed for protection. A third display was added later. That was Charlie Sternberg's last official work at Dinosaur Park. Sternberg passed away in 1981, age 96, after a long and distinguished career.

According to Dinosaur Provincial Park's second warden, G.F. Jerry Tranter (who succeeded Fowler in 1965), "The park might never have been had it not been for the determination and dedication of Roy Fowler." Tranter continued the Fowler tradition.

Fowler and the Brooks Kinsmen Club were responsible for the installation of the John Ware cabin museum at the park. Ware's original homestead furniture is on display. Local range brands and various samples of barbed wire are also preserved there.

After much controversy, in which locals wanted to keep the name Steveville or name the park after Dr. W.G. Anderson or the Sternberg family, a settlement was finally reached. January 1962 heralded the name change. The park, which was still undergoing road construction, became known as Dinosaur Provincial Park. The Steveville name was dropped to avoid confusion. (Visitors would lose their bearings trying to find a town that no longer existed.)

In the mid 1960s Dinosaur Provincial Park was still considered fairly inaccessible to tourists. The park is about 32 kilometres (20 miles) northeast of Brooks, off the Trans-Canada Highway, but people kept confusing it with Drumheller or even the Calgary zoo!

Newspapers of the day called the park "Canada's (or Alberta's) Best Kept Secret" or "the proverbial prophet without honour." It took years to develop political interest and proper accessibility. Dinosaur Provincial Park finally received its long overdue recognition when the federal government nominated the park to the United Nations Educational, Scientific and Cultural Organization's (UNESCO)

*Black rancher John Ware's cabin at the amphitheatre before more current renovations.*

## A World Heritage Site

World Heritage Site alumni in October 1979. World Heritage sites are established due to international concern for the lasting protection of significant, irreplaceable cultural and natural heritage values.

Recreation and Parks Minister Peter Trynchy (1979) commented, "The implications of Dinosaur's acceptance to the heritage list extends across all boundaries. The recognition of this park by the World Heritage Committee will do a great deal to advance its reputation as a site of scientific significance and remarkable beauty. Dinosaur Park has also gained renown for its rare and interesting natural terrain."

Dinosaur Provincial Park was officially declared a UNESCO World Heritage Site on June 19, 1980.

Parks Canada played a major role in implementing the World Heritage Convention for its inception in 1972. Other international sites of World Heritage status include: Egypt's Thebes, with its Necropolis; France's Roman Theatre; ancient cities of Damascus and Bosra; America's Yellowstone National Park and Grand Canyon National Park; and Canada's Head-Smashed-In Buffalo Jump. All have one thing in common with Dinosaur Provincial Park, "outstanding universal value."

The breathtaking badlands, the formidable fossil beds and striking geological variance were all overwhelming reasons why Dinosaur Provincial Park was the first site to be named to the World Heritage List under provincial or state jurisdiction.

*The Prairie Rattlesnake is a respected nomad not to be tampered with.* ~ *C. Wallis*

# Part Four

## Wildlife of Dinosaur Provincial Park

Visitors may be lured to Dinosaur Provincial Park by a fervent interest in the bronzed bones of dinosaur antiquity or the weird, chaotic shapes and forms of geologic strata, yet the Red Deer River valley supports a surprising array of natural beauty and is teeming with wildlife for all to see and enjoy. A drive to the park will unfold hints of what is in store. Watch for a startled sage grouse or a pronghorn running through a farmer's fence.

If it were possible to spend a year at the park, one might spot over 130 different species of birds. This includes migrants and winter residents.

The natural remoteness of the badlands leaves the area virtually undisturbed. The preserve region provides an ideal habitat for golden eagles. The eagles are sometimes seen soaring overhead like sentinels guarding the earth's inner secrets. The dark brown adult golden eagle has a wingspan of up to 2.3 metres (seven and a half feet). They are generally larger in all aspects than bald eagles, building huge nests in cliffs and preying on small mammals and other birds.

Swainson's hawks, marsh hawks, American kestrels, merlins (like sparrow hawks) and prairie falcons are the most commonly sighted raptors at Dinosaur Park. Gliding overhead, seeking a Richardson's ground squirrel (gopher), these and other hawks and falcons offer the budding naturalist a splendid chance to study each bird's discernible features. It is advisable to bring a good pair of binoculars.

*One of the most common sights at Dinosaur Provincial Park is the Cottontail Rabbit.*

*Prickly Pear Cactus in full, late spring bloom.*

Ferruginous hawks, rarely spotted, are on the endangered species list. Look for dark reddish-brown shapes under the adult in flight — those are its legs. The wings are more pointed and the tail longer than a golden eagle. Voles, field mice and other rodents are its mainstay.

Cliff dwellers include the ever-on-the-move rock wrens, with their pleasant variety of trills, calls and whistles; small, bluish-white cliff swallows, with their gourd-shaped mud nests; and house wrens, sparrow-size and grey with brownish flanks, who prefer taking refuge along the river's tree zone.

Evening welcomes the strange "haroom" (or "whoom") of the courting or insect-eating nighthawk as it dives into the inky blackness. The nighthawk (bullbat) unfolds its wings, plunging earthward, creating that haunting bovine bellow.

At dusk one might glimpse a great horned owl as it slowly ascends from an indistinct bluff to usher in the nocturn. Burrowing owls are found at the park. They actually live in the ground. When walking, their heads and tails bob up and down in a repetitious, comical style.

Coyotes wail and yelp in the distance. Though mainly nocturnal, these timid, scraggly beasts are occasionally active by day. Lengths vary up to 1.4 metres (54 inches), with long bushy tails. They hunt in pairs, often tracking down the abundant deer in winter months. Summer sends the coyote foraging for small rodents, fruits and berries. By today's standards coyotes are becoming endangered.

*Cottonwood trees, seen along the Cottonwood Trail, were used in burial ceremonies by the Blackfoot Indians (near the Red Deer River).*

*The Coyote is a shy creature. It is sometimes seen in the western, restricted area of Dinosaur Provincial Park.*

One only hopes the coyote's lively night serenade will never leave the baleful moon alone to rise solemnly over the rim of the badlands. They are not considered harmful to humans.

The heavens are a spectacular sight at Dinosaur Park on a clear, moonless evening. Winter nights might offer a surprising show of aurora borealis, but the summer sky serves up our finest meteor shower, the Perseids of August.

Park "bears" are the most unwelcome night marauders. This park nickname applies to skunks. Though not a serious problem, skunks do like garbage, so it is advisable to keep campgrounds litter-free for obvious reasons!

Brown bats are another denizen of the twilight. Though seldom seen, they are not dangerous. Brown bats feed on flying insects. Contrary to rumours, they will not attempt to tangle in hair and generally prefer to be left alone.

Dawn is greeted by delightful songbirds. Bluebirds flit by the Little Sandhill Creek and melodious meadowlarks careen from tree to fencepost and back again.

The Cottonwood Trail at Dinosaur Park probably provides the best route to indulge in birdwatching – Cedar waxwings, orioles, mourning doves, Say's phoebe, yellow warblers, you name it!

Avocets (like sandpipers) with upturned bills differ from willets, marbled godwits and long-billed curlews. It is advisable to bring a good bird reference book in order to determine the subtle differences. In flight just above prairie level, near marshes, these birds are fascinating to observe.

Black Hill's and Nuttall's cottontail and jackrabbits (white-tailed prairie hares) are extremely common at the park. Some hares grow to startling sizes. Watch for "camouflaged" rabbits as they blend discreetly along roadsides.

The deer population at Dinosaur Park is phenomenal. A visitor would be most unfortunate not to see a white-tail or mule-tail deer at one time or another.

Mule deer ("Jumping Deer") grow to two metres (six and a half feet) in length, are reddish in summer and brown-grey in winter. Known chiefly for their large ears and jumping gait, the does (females) can be found nursing their fawns during the early

*Mule Deer are often as curious as the visitors to the park.*

*Deserted eagle's nest at sunset. Notice the large dinosaur bone partially embedded at bottom of rills.*

*The majestic Golden Eagle may occasionally be spotted along the crags and cliff walls of eroded valleys.* ~ Cleve Wershler

*Bluebells are a common sight along the hiking trails at Dinosaur Provincial Park.* ~ *C. Wallis*

*Scorpions, though rarely seen, are indigenous to the park area and are extremely poisonous.* ~ *C. Wallis*

summer. Sometimes triplets are born, but usually it is sets of twins. These gregarious animals band together in winter. They outnumber the white-tail deer ten to one.

White-tails are tawny in summer and blue-grey in winter. Their coats are brighter than the mule deer. Ever alert, the white-tail will raise its tail to show its white rump when in retreat.

Pronghorns (resembling antelope) are generally seen en route to the park entrance, up on prairie level.

The long-tailed weasel, a solitary climber, will fight when cornered. This wary predator is brown in summer and grows up to 58 centimetres (23 inches) long. Weekend visitors will not likely spot this one!

The towering cottonwoods along the Red Deer River constitute the new Cottonwood (Flats) Trail. The cottonwoods, members of the poplar family, stand like ancient guardians protecting their kindred river life. They grow in both sexes to heights of 43 metres (140 feet), reaching diameters of 1.5 metres (five feet) and live 200 years or more. An understorey of water birch, aspen and willows support the cottonwoods. In late June cottonwood fluffs circulate in the breeze. These are transitory cottonwood seeds.

The main threat to the cottonwoods is the beaver. They are usually sighted during the dusk or dawn hours. Park trees are actually wired down in places to keep them safe from this furry, bark-gnawing mammal. The beaver is fearless, but will never attack.

Bobcats have been glimpsed by early morning fossil-hunting parties, but are very timid and rarely seen by visitors.

Plains grizzlies and mountain lions roamed the badlands when Prince Maximilian journeyed through the northwestern United States, along the Missouri River, in the 1850s, but these fearsome predators have long since departed for the protection of the Rockies.

The honking flight formation of Canada geese at Dinosaur Park is matched only by the sighting of a gliding great blue heron. The herons arrive in late March or April. Their young are fed on an inlet east of Steveville and are ready to leave the nest by August. Then the colony is deserted. In August herons can be found sunning themselves or fishing at the riverside.

Visitors often wonder if there are any poisonous snakes at the park. The prairie rattlesnake is the only one. Light grey-green with lozenge-shaped saddles along the middle of its back, this typical pit viper is a descendant of Triassic and Jurassic reptiles. Reaching lengths of up to 1.2 metres (four feet), this snake senses heat vibrations and will strike accurately, especially at night. Attacks on humans are rare. The rattlesnake only strikes if startled or cornered. In the spring, during skin-shedding, they become temporarily blind. Do not touch these snakes, better to report them to a park ranger.

All the ranger staff have standard first aid certificates. Emergency medical assistance and ropes for people who get stuck on ledges or precipices are always ready at ranger headquarters.

Scorpions (nocturnal) are rare, but do not make a habit of overturning rocks. Black widow spiders are also poisonous. These spiders spin their delicate webs in pipe openings and sinkholes, among other places. Black widow venom is 50 to 100 times more potent than rattlesnake venom — fortunately one does not get as much of it. These creatures are rarely seen.

So far there have been no reported snake bites, scorpion stings or spider bites at Dinosaur Park.

Bullsnakes are not venomous. Their hissing sound is only a protective measure. One old wives' tale had it that the bullsnake (gopher snake) would mate with the smaller rattlesnake to create a super dangerous breed. This is, of course, erroneous. In fact, some consider the bullsnake to be a natural enemy of the rattlesnake. Bullsnakes, like rattlers, devour mice and rabbits, but the bullsnake suffocates its prey, squeezing like a boa constrictor.

Clusters of flowers in alpine meadows are not the case in the badlands; this is due to a lack of sufficient precipitation. Gumweed (like a dandelion) grows along roadsides. Tea made from its flower heads and leaves is supposed to be good for a sore throat.

Wildflowers include prairie crocus, wild rose (Alberta's provincial flower), smooth Bluebeard's tongue and the Colorado rubber weed.

Horned lizards like Dakota toads and plains spadefoot toads, as well as several species of tiger beetles and "hairy" caterpillars, may be spotted scurrying over sandstone sediments in late June.

June is the month the beautiful prickly pear cactus and ball cactus (pin cushion) dot the badlands with their decorative blooms. Proper hiking footwear is advised. Cactus needles have a nasty habit of showing up when least expected. They are sometimes deemed the park's "most dangerous critter". Prickly pear cacti are in abundance. The pin cushion is harder to find, but its violet-pinkish flowers are striking compared to the waxy-yellow of the prickly pear.

Pediment and rill-pocked coulees greet the hiker with a sparse array of sagebrush, greasewood, juniper wood, prairie buttercups, moss phlox and lichen. The variance in vegetation depends on southern (dry) or northern (moist) climatic exposure in the badlands.

Nodding onions, wild prairie onions, prairie parsley, and particularly sage scintillate the senses with pleasant aromas to offset the barren scenery.

Between the lofty cottonwoods and the five varieties of sage, there is a transitional area in the park covered with thickets of chokecherry, currant, bull berry, buffalo berry and Saskatoon berry. The latter shrub has a good pie-producing berry in autumn, and the yellow buffalo berry is excellent for jam preserves, whereas the black chokecherry berries are astringent to the taste. Note that the chokecherry fruit stones are considered poisonous.

Near the viewpoints (at prairie level) or atop a distant, "lonesome" mesa, the prairie short grasses consist of blue grama, wheat, tufted-hair and June grass.

Amidst the ravaged sandstone cleavages, sand grass, foxtail barley, vetches, bladder fern, low whitlowwort, broom rape, Indian breadroot, sneezeweed and owl clover subsist on little or no moisture year round.

The badland's butterflies, including the shasta blue and Riding's satyr, complement the variegation of contrasting natural phenomena at Dinosaur Park.

Park naturalists are adept at pointing out the various flora and fauna, but plant or animal guidebooks come in handy. No wonder visitors return so often, Dinosaur Provincial Park offers something new and exhilarating on each visit.

*A contemporary oasis from the heat and the dust of surrounding badlands, Dinosaur Service Centre offers dining and restroom facilities for the weary visitor.*

# Part Five

## Dinosaur Provincial Park

There has been some debate over how tourists should get to Dinosaur Provincial Park. It was not until 1969 that access via the Trans-Canada Highway was available. Driving the 48 kilometres (30 miles) northeast of Brooks is the best route. Little brown provincial park signposts alert the driver of the necessary turns in the road, and, the grade is tolerable enough. The last major turn is north to the town of Patricia, 15 kilometres (nine miles) southwest of the park. The road into the park is known as Secondary Road 551.

Patricia has a hotel if the park happens to be full for the weekend. The hotel is clean and reasonable. On busy weekends, like the May long weekend, the park is usually full by Thursday afternoon. There are presently 44 designated campsites, with an overflow capable of handling an additional 56 units.

Drumheller, near where the new Tyrrell Museum is located, is about 140 kilometres (85 miles) northwest of Dinosaur Park. A Highway 10 exit, leading north off the westbound Trans-Canada Highway, will take visitors up to the Drumheller area. Otherwise one must almost go back to Calgary to find an alternate northerly route.

Experienced campers are generally prepared, but the one- or two-day visitor should bring lots of drinking water or cool beverages. The park's water is acceptable but high in iron content and may not be to all tastes. Binoculars, cameras and mosquito

*Two points of interest to look for on the bus tour.*
*Above - Camel foreground and pyramid in background, are pet names for these unusual rock formations in The Valley of the Castles.*

*Below - Stone Ti Pi*

repellent are suggested. Hiking boots and some kind of hat would also be a good idea. There is a small refreshment stand at the campgrounds.

The temperature in the summer can reach up to 49° Celsius (120° Fahrenheit) in the shade. Annual precipitation can be less than the Sahara Desert at times.

In 1966 the park registered 28,000 visitors. By 1985 that total had reached over 80,000 a year. It is wise to book tours and school visits well in advance.

The 8,900-hectare (22,000-acre) Dinosaur Provincial Park is 90% restricted from public access. Yellow "Natural Preserve" signposts throughout the park warn visitors not to trespass. The badlands constitute 200 square kilometres (7,000 square acres) of the park.

The Natural Preserve area is restricted in order to keep intact the delicate balance of nature within its boundaries.

A major emphasis is placed on park interpretive events, informing visitors to be aware of the park's fragility and the significance of the park's resources. The Natural Preserve is an integral part of Dinosaur Provincial Park, which is in keeping with the UNESCO World Heritage Site doctrines. Brochures are available at the park detailing the importance of conservation measures.

Hiking and bus tours are available to the public on a daily basis from May 1st to the Labour Day weekend. Guides accompany hikers on designated tours. The park is patrolled regularly.

Once inside the park, bus tours are best booked in advance, because tickets go fast. The four-kilometre (two-and-a-half-mile) tour through Deadlodge Canyon gives tourists a chance to see Mesozoic clam shells, the spectacular badlands formations and Phred, the 250-year-old buffalo skeleton, who, legend has it, met his demise seeking shelter from a storm. A caprock undoubtedly fell on him.

In one 1930s Alberta newspaper, the badlands were described as "great rugged mounds of clay . . . veined with horizontal layers of reddish-brown bog iron and purplish coal deposits." Famed Group of Seven painter A.Y. Jackson once called Dinosaur Provincial Park "the most paintable landscape in western Canada."

In those days imaginative names for strange park rock structures included "The Bee Hive," "The Devil's Table" and "The Sugar Loaf." Today the names are equally peculiar, in keeping with the weird scenes surrounding the visitor. The bus tour winds through the "Valley of the Castles," which abounds in hoodoo formations. A sphinx or camel-like structure stands solemn guard over a pyramid look-alike! One might think the Egyptian gods formed the park. The "Valley of the Moon" and the "Valley of the Gold" are not far away.

Valley of the Gold owes its name to a series of layers of overbank deposits. Silt, sand and some clay deposits were laid down beside river channel deposits in the flood plains of the rivers. Iron oxide is spread thinly throughout, and when the golden rays of the setting sun hit these layers, it becomes a glorious spectacle.

*A unique grouping of toadstool-like hoodoos.*

*A general overview of badland topography. Note the hoodoos, caprocks, rills, rock walls and pediments.*

*An evolving hoodoo in the midday sun.*

The local ranchers probably originated such coulee names as "Waterfall Coulee," "Irishman Coulee" and "Son-of-a-Bitch Coulee." Park staff from one decade to another named the badlands sites. A park naturalist once stated, "People tell us they see everything from elephants to Dolly Parton!"

One of the last edifices to be seen on the bus tour is the amazing eroded siltstone formation known as "the Petrified Tepee" or "Stone Tepee" (tipi).

Those fortunate enough to hike in the Natural Preserve will see "The Cathedral," trek over actual wagon ruts from Brown and Sternberg's dinosaur-hunting days, and eventually see "The Citadel," which towers serenely over the Centrosaurus bone bed.

The badlands have attracted their share of distinguished media visitors: "Companions of the Daughter . . . " from South America, David Suzuki filmed a dinosaur special on *The Nature of Things* for CBC, Ned Kelly produced a series called *The Making of a Continent* for the BBC, and segments of the 1983 film *Quest For Fire* were filmed within the park boundaries.

The Natural Preserve houses a vast assortment of skeletal remains, with many more exposed annually. Close to 500 skeletons have been distributed the world over in 35 institutions.

"It's good advertising for Alberta. We would love to get some of it back," says Dr. Currie of the Tyrrell Museum, "especially some of the rarer ones, but they put the money into it. If they hadn't done the work, things would have eroded and been gone now. Alberta wouldn't have the fame it has now. We wouldn't have the money right now to build

*An exhilarating aerial view of Red Deer River Valley.*

this museum, because nobody would be interested in dinosaurs. The key reason why we should not be upset is that we have still got all this stuff here. We're going to surpass all their collections within ten years or more."

Only Edmonton's Recreation and Parks Department will issue excavation permits, and these are strictly controlled. The Tyrrell Museum now has a direct say on where any specimen will end up after removal from the park. Prior to Tyrrell, everything was sent up to the museum at Edmonton.

It is to be clearly understood that anyone caught digging without a permit will have severe fines levied. Unfortunately paleontologists are still plagued by amateur collectors who unlawfully remove fossil material.

The sale of fossils is not a criminal offence in Alberta. Scientists fear that valued, unique specimens could be lost if random areas are not patrolled. This is a prime reason why Dinosaur Provincial Park protects its fossil beds.

Dr. Loris S. Russell recalls the bone bed which was desecrated near the Bleriot (Munson) Ferry — not at the park and not protected: "In 1929 I took a group of undergraduate students from Princeton there. One of them came away with the sacrum of a dinosaur (a piece about two feet long, weighing 50 pounds). We asked him what he was going to do with it. He said he was going to make a doorstop!

"Unfortunately people got in there two or three years ago and bulldozed the thing out. It was quite illegal, but the authorities didn't react to the word passed on by the locals. It was pretty well wiped out. Too late to do much now. They were collecting these bones to cut up and sell as souvenirs."

*Struthiomimus - a small extinct indigenous predator, not unlike Velociraptor as seen in "Jurassic Park" the movie.*

*A life size replica of Styracosaurus welcomes all visitors to the entrance of the Dinosaur Provincial Park field station.*

*A rampaging Albertosaurus at the Royal Tyrrell Museum.*

*A wondrous display of stratigraphic levels in the badlands along Little Sandhill Creek.* ~ *Cleve Wershler*

If fossils are found, report them to a ranger or naturalist. A fossil-finders's certificate will be awarded if the find is significant.

Duckbill dinosaur displays are housed for all to observe at the turning point on the bus tour. Two lookout points, at the "upper" park entrance, present a tabletop view of the awe-inspiring Red Deer River valley.

Once past the entrance building (in the canyon), the visitor may examine dinosaur footprints, housed displays of various fossilized skulls, bones and teeth, or peruse the interior of John Ware's cabin. The cabin holds pioneer relics of the wild west and information about park wildlife. There is ample parking and picnic grounds nearby.

An amphitheatre next to Ware's cabin is used by park naturalists to explain the wonders of the park. The bus tours depart from this point.

Interpretive programs are presented to inform the public about the park's natural heritage. One of these programs is "The Badlands at Twilight," which involves storytelling and the singing of songs about Dinosaur Park. It is a mood-setting program detailing the park's history. A feeling of mystery and love for the park is instilled. Other evening programs include photography and natural history hikes.

There are three major hiking trails at the park, with a fourth in the planning stages. The Cottonwoods Trail winds north of the campgrounds. Look for the brochure provided. Another trail slopes over the badlands just east of the campsites. Parking is provided, descriptive booklets are in boxes on the trailhead and a guest list is there to be signed. Signatures from all over the world are collected there.

*The Badlands Trail is one of the best hiking trails of the Park.*

The 2.5-kilometre (one-and-a-half-mile) Badlands Trail offers the visitor 16 station by station checkpoints. Each station describes specific geological points of interest. The trail is approximately a 45-minute walk through the heart of the badlands.

A lesser-known trail edges south behind the John Ware cabin. It takes the adventurous hiker up over ledges and steep hills to a refreshing view of the Little Sandhill Creek. This trail is not recommended for children or those afraid of heights.

The campsites are well organized, individually numbered lots, with grass, trees for shade and gravel paths. Government land (100 hectares or 250 acres) west of the park headquarters building, along the Red Deer River, will eventually become new campgrounds. Alcoholic beverages are allowed at campsites. Wood is provided for campers' fire pits. There are also covered camp kitchens with wood stoves.

*The Milk River Field Station (4 hours S. of Dinosaur Provincial Park) welcomes U.S. visitors from the South into Alberta's rich Dinosaur Heritage.*

*The field station at the Royal Tyrrell Museum of Palaeontology gives you an overview of the important historic and geological aspects of Dinosaur Provincial Park.*

A playground area is provided for children. Waste receptacles, outhouses, a public telephone and water are also provided.

Though not known primarily as a recreational park, Dinosaur offers cross-country skiing and snow-shoeing in winter (if there is enough snow), limited canoeing, fishing (if you have a provincial fishing licence, the river offers walleye, northern pike and goldeye), swimming and, of course, hiking.

Dinosaur Provincial Park is undergoing a master-planning stage to accommodate future generations. Increases in park staffing and facilities are being studied. There may be a small theatre, office space for visitor services staff and a reception area.

*The Royal Tyrrell Museum, in co-operation with the park, is planning a 930-square-metre (10,000-square-foot) field research station, which may house an indoor museum. A major interpretive centre will probably be added on as an expansion of the research station.*

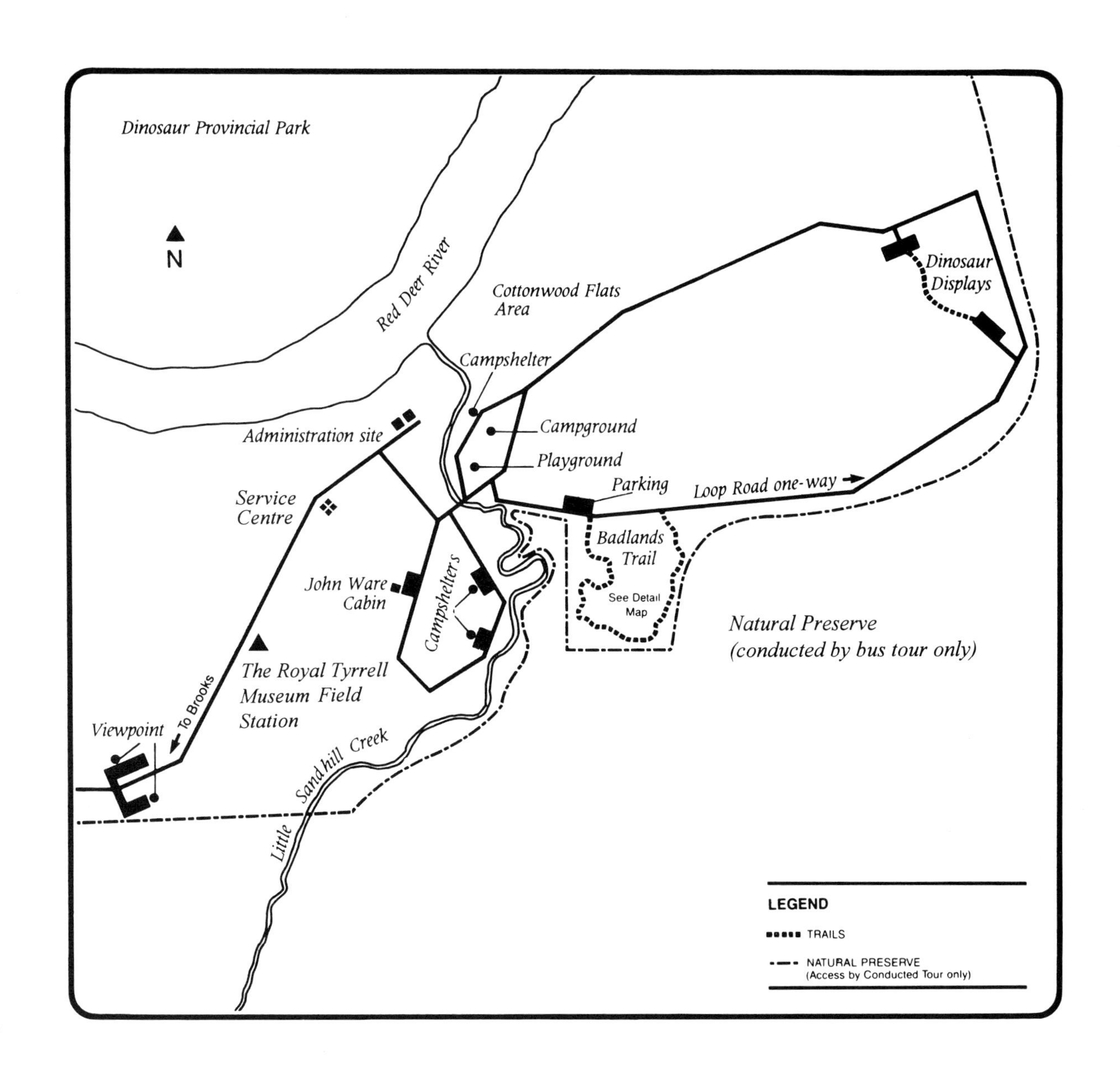


*The Royal Tyrrell Museum Field Station Page 60, Dinosaur Service Centre Page 50, Badlands Trail Page 59, John Ware Cabin Page 41.*

*A fine example of a clam shell bed discovery at Dinosaur Provincial Park.*

Dr. Phillip Currie, the chief vertebrate paleontologist of the new $28-million Tyrrell Museum complex, would like to see more thought put into the development of Dinosaur Provincial Park. "It's one of the few things in the province that will draw people away from Jasper and Banff. It is important to us that people know what we are doing is important, or a sense of excitement that what we are doing is interesting. The more people the better, but on the other hand we just have to make sure that it doesn't outstrip our ability to put people through the park in a meaningful way." Time stands still at the badlands, and more prehistoric wonders await.

Dr. Currie's long-range goal is to eventually know the park like the back of his hand.

*Main entrance to the park (as photographed in 1986). Note the petrified wood display in foreground.*

**Gordon Reid** spent three years photographing and researching ***Dinosaur Provincial Park***. He has since written a book on another World Heritage site, called ***Head-Smashed-In Buffalo Jump***, located near Fort MacLeod, Alberta. Both projects were supported with the generous assistance of the Royal Canadian Geographical Society. All photos in the text not otherwise noted are by the author.

**The Royal Tyrrell Museum of Paleontology** is located at Drumheller, Alberta. Many of the articulated remains of dinosaurs found at Dinosaur Provincial Park are on display there. For further information contact:

*Big Country Tourist Association*
*Box 2308*
*Drumheller, Alberta*
*T0J 0Y0*

For information about Dinosaur Provincial Park contact:

*South Alberta Travel Association*
*Box 605*
*Medicine Hat, Alberta*
*T1A 7G5*
*(403) 527-6422*

*The author.* *~ Photo courtesy of John Hampton*

# Biography

*Alberta native, Gordon Reid has written 3 books of a historical-geographical nature. They include* ***Poor Bloody Murder (Memoirs of the First World War)*** *(Mosaic Press, 1980),* ***Dinosaur Provincial Park*** *(1986) and* ***Head-Smashed-In Buffalo Jump*** *(1992). The latter two are Canadian World Heritage Sites.*

*Reid has also worked as a free-lance journalist with Oakville, Ont., papers and the Lethbridge Herald. In 1991, Reid wrote a ficticious script for an Alfred Hitchcock Mystery Board Game and has researched and interviewed many sports and motion picture related stars, over the years.*

*Reid resides in Oakville, Ontario and currently works for a publishing company based in Toronto.*

# Bibliography


Andrews, Roy Chapman. *All About Dinosaurs.* Random House 1953.

Bakker, Robert T. *The Dinosaur Heresies.* William Porrow & Co. 1986.

Carroll, R.L. *Vertebrate Paleontology*. W.H. Freeman & Co. 1988.

Charig, *A.J. A New Look At The Dinosaurs.* Heinemann, 1979.

Currie, Philip J. *The Flying Dinosaurs. Discovery Books*. Red Deer Press 1991.

Czerkas, Stephen & Sylvia. *Dinosaurs: A Complete World History.* Dragon's World Ltd. 1990.

Desmond, A.J. *The Hot-Blooded Dinosaurs*. Blond & Briggs, 1975.

Gardom, Tim. *The Natural History Museum Book of Dinosaurs.* Carlton Books, 1993.

Glut, Donald F. *The Dinosaur Dictionary*. Citadel Press, 1982.

Horner, John R. *Digging Dinosaurs*. Harper & Row, 1990.

Hsu, Kenneth J. *The Great Dying.* Ballantine, 1986.

Lambert, D. *Dinosaur Data Book*. Facts on File, 1990.

Lambert, David. *A Field Guide to Dinosaurs*. Avon Books, 1983.

Noble, Brian & Glenn Rollans. *Alberta, The Badlands*. McClelland & Stewart, 1981.

Paul, Gregory S. *Predatory Dinosaurs of the World.* Simon & Schuster, 1988.

Ratkevich, Ronald Paul. *Dinosaurs of the Southwest.* University of New Mexico Press, 1976.

Reid, M. & J. Sovak. *The Last Great Dinosaurs.* Discovery Books, Red Deer College Press, 1990.

Rogers, Katherine. *The Sternberg Fossil Hunters-A Dinosaur Dynasty*. Mounain Press, 1991.

Russell, Dale A. *A Vanished World: The Dinosaurs of Western Canada.* Nat. Museums of Canada, 1977.

Russell, D.A. *An Odyssey in Time: The Dinosaurs of North America*. University of Toronto Press & Norwood Press, 1989.

Spalding, David A.E. *A Nature Guide to Alberta* Hurtig Publishing, 1980.

Sternberg, Charles H. *Hunting Dinosaurs in the Badlands of Alberta, Canada*. 1917/1932

Stewart, Ron. *Dinosaurs of the West.* Lone Pine Publishers, 1988.

Thompson, Ida. *The Audobon Society Field Guide to North American Fossils*. Alfred A. Knopf, 1981.

Wallace, Joseph. *The Rise and Fall of the Dinosaur.* Friedman Group, 1987.

Wellnhofer, P. *The Illustrated Encyclopedia of Pterosaurs*. Salamander Books, 1989.